AF671717

THE SCIENCE OF VITAMIN E

LA FONCEUR

Eb
emerald books

emerald books

Copyright © 2024 La Fonceur

All rights reserved.

No part of this publication may be reproduced, stored in a retrieval system or transmitted in any form or by any means, electronic, mechanical, photocopying, recording or otherwise, without prior permission of author.

This book has been published with all efforts taken to make the material error-free. The information on this book is not intended or implied to be a substitute for diagnosis, prognosis, treatment, prescription, and/or dietary advice from a licensed health professional. Author and publisher don't assume and hereby disclaim any liability to any party for any loss, damage, or disruption caused by errors or omissions, whether such errors or omissions result from negligence, accident, or any other cause.

While every effort has been made to avoid any mistake or omission, this publication is being sold on the condition and understanding that neither the author nor the publishers or printers would be liable in any manner to any person by reason of any mistake or omission in this publication or for any action taken or omitted to be taken or advice rendered or accepted on the basis of this work. Some contents that are available in electronic books may not be available in print, or vice versa.

CONTENTS

Introduction **5**

Chapter 1: Basics of Vitamins **7**

Chapter 2: Everything You Need to Know About Vitamin E **15**

Chapter 3: Importance of Vitamin E **23**

Chapter 4: 10 Richest Food Sources of Vitamin E **36**

Chapter 5: Potentially Dangerous Vitamin E Combinations **48**

Chapter 6: Healthy Vitamin E Combinations **51**

Chapter 7: Diet Plan **56**

Chapter 8: Recipes **58**

Stir Fried Broccoli ***58***

Peas Avocado Soup ***60***

Kiwi Smoothie ***62***

Almond Avocado Pudding ***63***

References **65**

About the Author **73**

Read More from La Fonceur 74

Connect with La Fonceur 74

INTRODUCTION

The term "vitamin" is often thrown around. While we may have some knowledge regarding vitamins, do we really know all about vitamins? While we may have some vitamin knowledge, our awareness is often limited to what we hear from health advocates or simply dietary supplement manufacturing companies. You may come across countless articles promoting the benefits of vitamins, and they often conclude with recommendations for specific supplements.

When questioned about the benefits of Vitamin E, the usual response from most individuals would be Vitamis E is good for skin and blood thinning. However, these Vitamin E does much more than that. Consuming it through natural sources can provide various other health benefits that may surprise you. In fact, Vitamin E has been known to affect your health in numerous ways, some of which are yet to be fully discovered. This is why relying on vitamin supplements may not provide the same results that can be effortlessly obtained through natural sources. If you're not focusing on getting your necessary vitamins from food, you're missing out on a lot of potential health benefits.

Get all your answers about Vitamin E with ***The Science of Vitamin E*** book. Learn about its crucial role in maintaining good health and the latest scientific findings and how these can affect your vitamin decisions. Clear up common vitamin-related dilemmas, such as how to tell if you're deficient in vitamin E and when to get tested.

Learn about the advantages of combining Vitamin E with other vitamins and minerals for optimal health benefits, as well as the potential consequences of taking Vitamin E with particular foods or medications. This guide covers both beneficial and harmful combinations of Vitamin E.

Furthermore, learn about nutrient-rich vegetarian options that are high in Vitamin E. By consuming these foods, you can avoid Vitamin E deficiencies and maintain good overall health, reducing the likelihood of infections and chronic illnesses such as cancer, diabetes, high blood pressure, and cognitive decline. Plus, explore some nutritious and easy-to-cook vegetarian recipes that can be included in your diet to maximize the health benefits of Vitamin E.

CHAPTER 1

BASICS OF VITAMINS

Vitamins are organic compounds required by the body in small quantities to perform various normal functions Vitamins can be essential or non-essential. Essential nutrients are crucial for the normal function of the body, and the body cannot produce them, so they must be obtained through food.

Vitamins differ from macronutrients such as carbohydrates, proteins, and fats because they do not provide energy and are required in smaller quantities. They are called micronutrients because they are needed in small amounts, but this does not make them any less important than macronutrients.

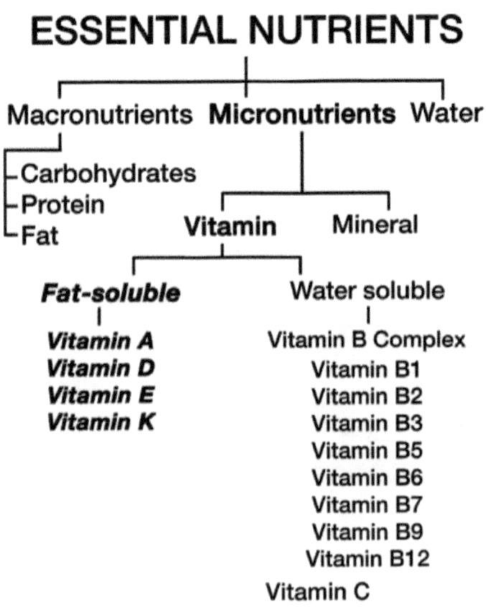

While vitamins are now a topic of general discussion, it may surprise you that they were discovered not so long ago. In fact, all known vitamins were identified during the period between 1912 and 1948.

CLASSIFICATION OF ESSENTIAL VITAMINS

There are 13 essential vitamins that are classified as water-soluble vitamins and fat-soluble vitamins.

Water-Soluble Vitamins

Water-soluble Vitamins are Vitamin B1, B2, B3, B4, B5, B6, B7, B12 and Vitamin C. These vitamins are called water-soluble as they dissolve in water. There are nine of them, including Vitamin B1 to B12 and Vitamin C. Since they dissolve in water, they are easily absorbed, and any excess is excreted in urine quickly without being stored in the body. This is why it is essential to consume water-soluble vitamins regularly to maintain adequate levels.

Fat-Soluble Vitamins

Fat-soluble vitamins are Vitamin A, Vitamin D, Vitamin E and Vitamin K. Fat-soluble vitamins are not soluble in water but instead, dissolve in fats. There are five vitamins that are classified as fat-soluble vitamins - vitamins A, D, E, and K, and these are absorbed by the body in a similar way to dietary fats. When you consume fat-soluble vitamins with healthy fats, their absorption in the intestine increases. Unlike water-soluble vitamins, fat-soluble vitamins are stored in the body and can be used whenever the body requires them. The body stores fat-soluble vitamins in the liver, muscles, and fatty tissue (adipocytes). This means that a continuous supply of fat-soluble vitamins is not necessary, as they are not excreted from the body as quickly as water-soluble vitamins. However, consuming adequate amounts is important to reach the daily recommended intake.

Too much of these vitamins can lead to toxicity (Hypervitaminosis), especially for vitamin A from animal sources (retinol) and vitamin D. Toxicity usually occurs when taking supplements, not from food. A balanced diet typically provides sufficient amounts of fat-soluble vitamins.

Types of Vitamin Deficiency

There are two types of vitamin deficiency:

Primary deficiency

Secondary deficiency

Primary deficiency

Primary deficiency can happen when you do not eat enough foods that are high in vitamins.

Reasons:

- Poor diet.
- Unavailability of a particular vitamin-rich food in that region.

Secondary deficiency

A secondary deficiency can happen when the body cannot properly absorb or use vitamins.

Reasons:

- Poor lifestyle choices such as smoking and drinking alcohol.
- Use of medications that interfere with the absorption of vitamins.
- An underlying disorder that limits the absorption of vitamins.

OXIDATIVE STRESS , ANTIOXIDANTS AND INFLAMMATION

Let's first understand some important terms before getting deep into Vitamin E.

Oxidative Stress and Antioxidants

Oxidative processes are normal in the body and are important to provide energy for many cellular functions. Oxygen is used by cells to generate energy, with free radicals being produced as a by-product. In small amounts, free radicals are beneficial as they can kill pathogens and regulate cell growth and death. However, excessive free radicals can cause chronic diseases. The body produces free radicals during normal cell metabolism, but external sources like radiation, pollution, cigarette smoke, and medication can also expose you to these harmful free radicals. Drinking alcohol and consuming excess sugar, and fat may also contribute to free radical production.

Free radicals are unpaired electrons that like to be paired. Antioxidants are produced by the body to balance free radicals and stop them from causing damage. Oxidative stress occurs when there is an imbalance between free radicals and antioxidants, and the accumulation of free radicals in the body cannot be gradually eliminated. This process causes unpaired free radicals to pair with electrons in fat tissue, proteins, and DNA, and this can damage cells and tissues, increasing the risk of developing chronic and degenerative diseases like cancer, rheumatoid arthritis, cataracts, diabetes, aging, as well as cardiovascular and neurodegenerative diseases.

Some foods have antioxidant properties and can delay or inhibit cellular damage by their free radical scavenging ability. These antioxidant foods are capable of breaking the free radical chain reaction by donating an electron to a free radical without destabilizing themselves. The three vitamins that act as antioxidants are pro-vitamin A (carotenoids), vitamin E, and vitamin C. Since these micronutrients cannot be produced by the body, they must be obtained through the diet.

Plant-Based Antioxidants

Nutrient

Vitamins: Vitamin A, Vitamin C and Vitamin E.

Minerals: Copper, zinc, selenium, and manganese.

Non-nutrient

Phytochemicals: Polyphenols such as flavonoids, lignans, and stilbenes.

Inflammation Anti-inflammatory Action

Let's first understand what is inflammation

Is inflammation good or bad?

Both. The process of inflammation is a natural defense mechanism of the body. It is the process by which the immune system identifies and eliminates harmful and foreign bodies, and begins the healing process. there are two types of inflammation, acute inflammation, which lasts for a few days, is helpful in cases where the body experiences injury or harm. For instance, if you have a cut on your finger, your body sends inflammatory cells to the injury site to start the healing process.

Chronic inflammation is a type of inflammation that occurs over a longer period of time and can last for several months or even years. Even when there is no external threat, your body continues to send inflammatory cells, which can create an ongoing and unnecessary inflammatory condition. This can eventually harm healthy tissues in the long run. Chronic inflammation is the primary cause of many chronic diseases, including rheumatoid arthritis, chronic obstructive pulmonary disease (COPD), diabetes, and cancer.

Chronic inflammation can be caused by several factors, including environmental pollutants, auto-immune diseases, infection, and untreated acute inflammation. Lifestyle factors, such as stress, obesity, alcohol consumption, and smoking, can also contribute to inflammation in the body. It is crucial to treat inflammation promptly, as not doing so can result in life-threatening consequences.

All fat-soluble vitamins A, D, E, and K are anti-inflammatory. Each one has a different pathway to reduce inflammation in the body.

CHAPTER 2

EVERYTHING YOU NEED TO KNOW ABOUT VITAMIN E

The next essential fat-soluble vitamin is Vitamin E. It actually consists of eight different chemical forms - four tocopherols and four tocotrienols. Each form has its own unique antioxidant properties. These tocopherols and tocotrienol forms are as follows:

Tocopherols: alpha-, beta-, gamma-, and delta-tocopherol.

Tocotrienol: alpha-, beta-, gamma-, and delta-tocotrienol

There are eight different forms of vitamin E, each with varying levels of biological activity. However, most of these forms are quickly metabolized by the liver and eliminated from the body. Out of all, only alpha-tocopherol maintains high blood and cellular concentrations, as only this form is re-secreted by the liver through the hepatic alpha-tocopherol transfer protein. This is why vitamin E is commonly referred to as tocopherol or alpha-tocopherol.

Vitamin E is essential for various bodily functions. It acts as an antioxidant that protects cells from harmful free radicals. In the next chapter, we will delve into the details of antioxidants, free radicals, and other crucial roles of Vitamin E. Let's take a look at the recommended daily amount of Vitamin E.

Recommended Intake of Vitamin E

According to the National Institutes of Health Office of Dietary Supplements, in order to fulfill your body's vitamin E (alpha-tocopherol) requirements, the recommended average daily intake amounts are as follows:

Age	Males	Females
0–6 months	4 mg	4 mg
7–12 months	5 mg	5 mg
1–3 years	6 mg	6 mg
4–8 years	7 mg	7 mg
9–13 years	11 mg	11 mg
14+ years	15 mg	15 mg

How common is vitamin E deficiency?

It is rare to experience Vitamin E deficiency due to a diet low in this vitamin. Usually, underlying conditions are the cause. When there is not enough fat present in the digestive tract, the body cannot efficiently absorb fat-soluble vitamins, such as Vitamin E. Individuals who follow extremely low-fat diets may have lower levels of Vitamin E. Additionally, those with fat-malabsorption disorders may also become deficient in Vitamin E. There are certain diseases that prevent fat absorption in the body, which can lead to irregularities in fat absorption or metabolism and ultimately result in Vitamin E deficiency.

Certain health conditions that can lead to vitamin E deficiency are:

- **Crohn's Disease:** Bile acids play a crucial role in facilitating the absorption and transportation of vitamin E. Crohn's disease tends to affect the distal part of the small intestine, known as the terminal ileum. Any damage to the terminal ileum can result in the inadequate absorption of bile acids, leading to vitamin E deficiency.

- **Cystic Fibrosis** is an inherited disorder that results in significant harm to the digestive system and the inability to secrete pancreatic enzymes for absorbing vitamin E.

- **Babies Born prematurely** and weighing less than 1500 grams may have a deficiency in vitamin E.

- **Genetics.**

- **Liver Disease.**

- **Abetalipoproteinemia:** A rare inherited disorder called abetalipoproteinemia results in poor absorption of vitamin E.

- **Ataxia** is another rare inherited disorder caused by a defective or absent alpha-tocopherol transfer protein in the liver, which leads to severe vitamin E deficiency.

What health problems can be caused by Vitamin E deficiency?

If you have a mild deficiency of vitamin E, you can rectify it by incorporating vitamin E-rich foods in your diet instead of taking supplements. However, if your deficiency is due to conditions like Crohn's disease, cystic fibrosis, or other illnesses, your doctor may recommend taking supplements based on your condition. Neglecting vitamin E deficiency may result in severe health issues, including:

Heart Problems: Severe Vitamin E deficiency may lead to the deterioration of muscles, including the heart muscle, and potentially result in cardiac failure.

Hemolytic Anemia: Vitamin E deficiency may lead to the oxidation of red blood cells and ultimately lead to their destruction or hemolysis. In this condition, the body breaks down red blood cells at a faster rate than it can produce them, causing a depletion of red blood cells, resulting in anemia.

Nerve Damage: Vitamin E prevents free radical damage. Nerve cells in the brain are vulnerable to the damaging effects of free radicals. Without enough Vitamin E, nerve cells can die off, which may result in loss of sensation in the extremities (peripheral neuropathy), slurred speech, and loss of reflexes in the legs

Retinopathy of Prematurity (ROP): Retinopathy of Prematurity (ROP) is a condition that affects premature infants, causing the growth of abnormal blood vessels in their eyes. Vitamin E plays a crucial role in protecting cell membranes from oxidation, preventing the oxidation of polyunsaturated fatty acids, which can contribute to the development of ROP. If a premature infant is deficient in vitamin E, they may be at greater risk of developing retinopathy of prematurity

---

While it may not be common to experience a vitamin E deficiency from a low intake of this nutrient in your diet, it is still important to consume enough vitamin E through natural food sources to ensure proper bodily function. Your body may exhibit signs that you are not getting enough vitamin E from your diet. Minor deficiencies can often be corrected through dietary adjustments rather than supplementation. However, it is important to be aware of moderate to severe symptoms of vitamin E deficiency.

Symptoms of Vitamin E Deficiency

If you notice sudden hair loss, unusual dry skin, or more frequent infections, it could be a sign that you are not consuming enough vitamin E in your diet. Other severe symptoms of vitamin E deficiency may include:

- Muscle weakness
- Vision problems
- Difficulty in walking
- Hair fall
- Dry and flaky skin
- Weakened immunity

When should I get tested to rule out a possible Vitamin E deficiency?

If you experience any of the symptoms mentioned above, it is recommended that you undergo a blood test to verify if you have a deficiency in vitamin E.

Am I taking too much Vitamin E?

Consuming too much vitamin E through food is unlikely. It's safe and recommended to have a balanced diet rich in vitamin E from fruits, vegetables, and whole grains. Taking vitamin E supplements without consulting a doctor is not recommended. Excessive intake of vitamin E supplements rather than obtaining it from food sources can lead to hypervitaminosis. Symptoms of excessive vitamin E intake may include:

Fatigue

Diarrhea

Nausea

It's important to be cautious about taking too much vitamin E through supplements, as it can lead to vitamin E toxicity and potentially life-threatening consequences. To determine any overdose of vitamin E, it is recommended that you consult with your doctor and have a vitamin E test done.

Diagnostic test to determine Vitamin E deficiency or overdose:

The Vitamin E test, also known as the alpha-tocopherol test, is a blood test that measures the level of Vitamin E in your bloodstream. The normal range for Vitamin E levels is:

Adults (18 Years – 150 Years): 5.5 – 17 mg/L

Children (0 – 18 Years): 3.8 – 18.4 mg/L

*Normal ranges may differ slightly between different laboratories.

Do I need to do any preparation before the test?

• This test is taken on an empty stomach, meaning you should not have food or drink for 12-14 hours before the test.

• Do not consume alcohol 24 hours before the test.

• Do not take vitamin supplements 24 hours before the test.

CHAPTER 3

IMPORTANCE OF VITAMIN E

ROLE OF VITAMIN E IN THE BODY

Antioxidant Activity

Vitamin E plays a crucial role as an antioxidant, protecting the body from free radical damage. It is one of the most potent antioxidants available, with the ability to scavenge free radicals and prevent damage to cells. Vitamin E works in two ways to prevent free radical damage. Firstly, it provides a hydrogen atom to unpaired free radicals, breaking the chain reaction and preventing DNA, lipids, and protein damage. Secondly, it limits the production of free radicals altogether. Free radicals are highly reactive and can cause damage to cell structures.

ROS (reactive oxygen species) are formed as a by-product of the oxidation of dietary fats to produce energy in the body. In lesser amounts, ROS helps to kill invading pathogens and regulate normal physiological functions at the cellular level. However, excessive ROS formation can lead to oxidative stress and result in various diseases. Vitamin E, a fat-soluble antioxidant, helps prevent ROS production and thus protects against chronic diseases caused by free radicals.

Anti-Inflammatory Action

Inflammation is a defense mechanism used by the immune system to protect the body against foreign attackers, such as viruses and bacteria. Although inflammation is a beneficial process, it can sometimes harm the body. In some cases, the immune system mistakenly attacks the body's own cells, leading to harmful inflammation. This can result in autoimmune diseases like rheumatoid arthritis, osteoporosis, psoriasis, inflammatory bowel diseases, and type 1 diabetes. To reduce inflammation in the body, certain substances interfere with the chemical reactions that cause it. These substances are called anti-inflammatory agents. Vitamin E is one such agent that plays an important role in anti-inflammatory processes. COX-2 enzymes produce prostaglandins that promote pain and inflammation. Vitamin E inhibits these inflammation-causing COX-2 enzymes.

Vitamin E also suppresses the production of pro-inflammatory cytokines. Free radicals activate transcription factor NFκB, which leads to the production of inflammation-causing cytokines. Vitamin E neutralizes free radicals before they can activate NFκB, suppressing cytokine production and reducing inflammation.

Inhibits Platelet Aggregation

Platelets play a crucial role in hemostasis and thrombosis, which are processes related to blood clotting. Hemostasis is the initial stage of clotting, where the blood transforms from a liquid state to a gel-like state to prevent further bleeding. While blood clotting is vital in wound healing, it can be harmful if it occurs inside blood vessels. Thrombosis is the medical term for blood clot formation within a blood vessel. This can impede blood flow through the circulatory system, leading to health complications like heart attacks or strokes. Alpha-tocopherol can help inhibit protein kinase C activity, which is an enzyme involved in platelet secretion and aggregation.

Vitamin E has another way of preventing platelet aggregation: by increasing the production of vasodilators. Prostacyclin, a member of the prostaglandin family, is a powerful inhibitor of platelet activation and a vasodilator. Vitamin E, specifically D-alpha-tocopherol, triggers the release of prostacyclin from the endothelium cells lining the inside of blood vessels. This, in turn, dilates blood vessels and prevents platelet aggregation. This effect is due to vitamin E having an opposing effect on the two key regulatory enzymes in prostacyclin biosynthesis, which results in a net increase in the production of vasodilator prostacyclin in endothelial cells.

Role in Immune Enhancement

Direct Effect: Vitamin E is found in higher levels in immune cells compared to other cells in the body. It's actually one of the most effective nutrients when it comes to regulating immune function. Vitamin E has the ability to protect the polyunsaturated fatty acids (PUFAs) in the cell membrane from oxidative damage. Immune cells are highly influenced by cell membrane composition and structure as their membranes are a primary site for translating external signals. By preventing oxidation and cell membrane damage, Vitamin E helps maintain membrane integrity and signal transduction, which ultimately affects the function of immune cells.

Indirect Effect: T cells are the white blood cells of the immune system that has a central role in the immune response to infection. They carry out their function either as killer cells that kill cells infected with a virus and cancer cells or as helper cells, aiding B cells in producing antibodies. When T cells are not functioning properly, it can lead to a higher risk of contracting infectious diseases and a weaker response to immunization.

Vitamin E can regulate the immune system's T cells by modulating inflammation-causing cytokines and prostaglandin E2 (PGE2). PGE2 is a pro-inflammatory prostaglandin that suppresses T-cell response by inhibiting T-cell proliferation and Interleukin-2 (IL-2) production. However, Vitamin E can inhibit prostaglandin E2 production by suppressing cyclooxygenase 2 (COX2) enzyme activity that converts arachidonic acid to prostaglandins. This results in an increase in the cell division and Interleukin-2-producing capacity of naïve T cells, which in turn increases the percentage of T cells. IL-2, which is responsible for activating T cells, has the potential to kill cancerous cells and reduce the size of tumors wherever they develop in the body.

Top 10 Most Significant Health Benefits of Vitamin E

1. Prevent Coronary Heart Disease

When cholesterol builds up in the walls of arteries that supply blood to the heart, it can block blood flow and cause coronary heart disease. This build-up can cause the arteries to narrow over time. This process is known as atherosclerosis, and the cholesterol deposits are called atheroma. A crucial factor in atherosclerosis is low-density lipoprotein

(LDL) cholesterol. Vitamin E can effectively prevent or delay coronary heart disease by inhibiting the oxidation of LDL cholesterol. Additionally, vitamin E can help prevent blood clots that may lead to a heart attack by inhibiting platelet aggregation.

2. Prevent Cancer

Vitamin E plays an important role in protecting cell integrity from harmful free radicals, which can contribute to the development of cancer. A deficiency of vitamin E can increase the risk of cancer. Vitamin E prevents prostate cancer in men who are at high risk of prostate cancer. Consuming foods rich in vitamin E can greatly decrease the likelihood of developing advanced prostate cancer, particularly for individuals who have quit smoking.

Nitrosamines, which are metabolites of nicotine found in cigarette smoke, are major cancer-causing agents. These nitrosamines can generate free radicals that damage the DNA. Additionally, nitrosamines can also be created from nitrate and nitrite, which are added to processed meat to keep it fresh for longer. When processed meat is cooked at high temperatures, the protein in it reacts with nitrates to create an ideal environment for the formation of nitrosamines, which are known to be carcinogenic. Vitamin E can help block the formation of these cancer-causing nitrosamines and protect against various types of cancer by enhancing immune function.

You may be curious as to why nitrates are considered carcinogenic, even though nitrates in beetroot can help lower blood pressure and prevent heart disease (as specified in the book **Eat to Prevent and Control Disease**). The reason is that vegetables, like beetroot, are not high in protein and are rarely cooked at high temperatures. Additionally, vegetables contain protective elements such as vitamin C, fiber, and polyphenols, all of which have been shown to decrease the formation of nitrosamines. These factors contribute to the health benefits of beetroot and other vegetables despite the presence of nitrates.

3. Prevent Eyes Disorder

As you age, you may experience age-related eye problems such as Age-related macular degeneration (AMD) and cataracts. These issues are the most common causes of vision loss among older individuals. Your retina and macula are essential components of your eye that enable clear and central vision. Macular degeneration occurs when the macula degrades, which typically happens as you age, hence the term age-related macular degeneration. Cataracts, on the other hand, occur when protein accumulates in the lens of your eye, causing cloudy vision. Oxidative stress caused by free radicals can harm the macula and protein within the eye. Vitamin E functions as an antioxidant and can prevent these types of damage to the eyes. Numerous studies have shown that vitamin E obtained through a healthy diet is highly effective in preventing eye disorders, although taking vitamin E supplements has not yielded positive results.

4. Prevent Influenza

Influenza is a type of viral infection that affects your respiratory system, including your lungs, nose, and throat. While vitamin E does not offer specific antiviral benefits, its antioxidant properties can help protect your lungs and prevent oxidative damage caused by influenza.

Severe influenza can cause damage to the lungs by triggering viral replication and inflammation, which generates free radicals that harm the cellular membranes of blood vessels. These excess free radicals cause the oxidation of unsaturated fats in cell membranes. Since vitamin E is fat-soluble, it accumulates in fat membranes and exerts antioxidant action there. It reacts with unpaired electrons of free radicals and can prevent them from reacting with adjacent fatty acid side chains. Due to its effectiveness in preventing oxidative damage through its free-radical scavenging activity, vitamin E is the most efficient antioxidant for influenza virus infection.

5. Prevent Asthma

Research shows that individuals with asthma often have lower levels of vitamin E, and vitamin E deficiency is linked with the severity of asthma symptoms. Vitamin E can reduce inflammation and act as an antioxidant, which could be beneficial for those with asthma. However, there have been conflicting results from different studies on the effects of vitamin E supplements on asthma. Some studies suggest that high levels of alpha-tocopherol, a major form of vitamin E, could lead to a decrease in lung function, whereas other studies dispute this finding and instead propose that alpha-tocopherol can actually improve lung function, while gamma-tocopherol might reduce it.

Asthmatics have low levels of vitamin E, which is essential in reducing inflammation. Therefore, increasing dietary intake of vitamin E may help prevent or control allergic diseases and asthma. Furthermore, vitamin E has been shown to reduce levels of mucins, which affect the stickiness of mucus. This is particularly important as people with asthma often have elevated levels of mucins.

6. Prevent Osteoporosis

Osteoporosis is a bone disease characterized by low bone mass and an increased risk of fractures. Normally, your body constantly replaces bone tissue to keep your bones healthy. However, in osteoporosis, new bone is not formed as old bone is removed, resulting in bone loss. This loss of bone mass weakens the bones, making them more susceptible to fractures. Chronic inflammation is the primary cause of osteoporosis. When inflammation occurs near the bones, it often increases bone resorption, degrading the bone without any subsequent coupling to new bone formation. Older people are particularly prone to osteoporosis, as they tend to produce more pro-inflammatory cytokines as they age.

Inflammation is triggered when free radicals activate transcription factor NFκB, which leads to the production of the cytokines interleukin-1 and interleukin-6. These pro-inflammatory cytokines cause bone resorption. Vitamin E can help prevent osteoporosis by scavenging and neutralizing free radicals before they can activate transcription factor NFκB. Vitamin E also enhances the internal antioxidative enzymes within the bone, which further help prevent the activation of NFκB.

In addition, vitamin E acts as an anti-inflammatory agent to safeguard bones from degradation. Vitamin E decreases the production of prostaglandins, the culprits behind inflammation. It hinders the activity of COX-2 enzymes responsible for the production of prostaglandins.

7. Prevent Alzheimer's Disease and Other Neurodegenerative Diseases

Neurodegenerative diseases are irreversible conditions that primarily affect nerve cells in the human brain. This happens when nerve cells lose their structure or function, leading to cell death. Over time, free radical damage to nerve cells contributes to the development of these diseases. In particular, oxidative stress is a major player in Alzheimer's disease and is involved in its initiation and progression. The brain is composed of 60% fat and consumes 20% of the total oxygen in the body. Lipid peroxidation and protein oxidation in the brain can trigger Alzheimer's disease. Vitamin E can reduce oxidative stress and improve memory and cognitive deficits.

Research shows that consuming more vitamin E through a balanced diet, rather than relying on high-dose supplements, is linked to a lower risk of neurodegenerative diseases. You may significantly reduce your chances of developing neurodegenerative conditions by incorporating vitamin E-rich foods into your diet, along with other potent antioxidants like carotenoids and vitamin C.

8. Enhance Skin Health

Exposure to ultraviolet rays from sunlight, dust, air pollution, and smoke can cause free radicals to form in your skin cells, leading to brown spots, wrinkles, and premature aging. Applying vitamin E oil topically may help reverse this damage by acting as a free-radical scavenger. Additionally, using vitamin E and vitamin C together can provide even greater protection against UV rays, as vitamin C enhances vitamin E's antioxidant action and produces synergistic results.

9. Boost Hair Health

Vitamin E is essential in maintaining your scalp health. Vitamin E is an antioxidant that helps protect the cellular membrane of your hair follicles from damage-causing free radicals. This action ensures that oxygen and nutrient-rich blood are delivered to your hair, which helps moisturize and hydrate dry and brittle hair. This, in turn, helps strengthen the hair follicles and prevents hair loss.

Additionally, vitamin E has anti-inflammatory properties that help reduce inflammation that can cause hair follicle cells to break down. This helps prevent hair loss and promotes healthy and strong hair growth.

Vitamin E also helps prevent premature graying of hair and split ends. Free radicals can damage cells and accelerate the aging process, but vitamin E helps prevent the depletion of cells, which helps combat the premature graying of hair. Moreover, oxidative damage to the hair follicles can cause split ends, but applying vitamin E oil topically can help seal them and prevent further splits.

10. Premenstrual Syndrome (PMS)

Approximately 90% of women experience premenstrual syndrome (PMS) symptoms, such as pain, mood swings, stress, bloating, fatigue, and other physical and emotional changes, one to two weeks before their period begins. These symptoms can adversely affect the quality of life and daily work. Fortunately, antioxidant vitamin E can help alleviate PMS symptoms by reducing lipid oxidation and inhibiting the release of arachidonic acid and its conversion to pain-producing prostaglandins. This makes vitamin E significant in reducing the severity and duration of menstrual cramps. Additionally, vitamin E can significantly reduce irritability, stress, and other mood symptoms associated with PMS.

CHAPTER 4

10 RICHEST FOOD SOURCES OF VITAMIN E

NUTS & SEEDS

Nuts and seeds are the primary sources of vitamin E due to their abundance of healthy fats and numerous health benefits. Incorporating these into your diet is a wise investment in your overall health. Consuming a handful of mixed nuts and seeds daily can reduce your risk of chronic diseases. However, during summer, limiting consumption to four times a week is advised, as nuts and seeds can generate heat in the body and potentially cause mouth ulcers.

While most dry fruits contain vitamin E, some are richer sources than others. Let's take a look at the ones with the highest vitamin E content:

1. Sunflower Seeds

Sunflower seeds are a rich source of vitamin E. Just a handful (about 30 grams) of dry roasted sunflower seeds contain 49% of your daily recommended intake of vitamin E. These seeds also contain other beneficial nutrients like potassium, magnesium, fiber, and zinc that can help lower the risk of developing high blood pressure, diabetes, and heart disease. Additionally, the vitamin E, flavonoids, and plant compounds in sunflower seeds can help reduce chronic inflammation.

However, if you suffer from arthritis, it's best to avoid consuming excessive amounts of sunflower seeds because sunflower seeds contain high levels of omega-6 fats, which can aggravate your arthritis symptoms. It's important to consume omega-6 fatty acids in moderation.

2. Almonds

Almonds are not only delicious but also nutritious. They contain a variety of healthy components such as vitamin E, fiber, monounsaturated fats, and protein. Additionally, they are rich in essential minerals like magnesium, iron, calcium, copper, and zinc. Eating a handful of dry roasted almonds (around 23 kernels) can provide you with 45% of your recommended daily vitamin E intake. Almonds are a great snack option because they are low in carbohydrates and high in protein. The protein in almonds slows down digestion and makes you feel fuller for longer, aiding in weight loss by reducing hunger. Moreover, eating almonds can help lower the risk of type 2 diabetes and hypertension, as they are high in magnesium. Vitamin E in almond oil can also be beneficial for your skin, healing sun damage and giving you a smooth and glowing complexion. And for those looking to improve hair growth, massaging your scalp with almond oil can provide numerous health benefits.

For optimum health benefits, consider consuming soaked almonds. The skin of almonds contains phytic acid that can hinder the absorption of vital minerals like iron, magnesium, zinc, copper, and calcium. Soaking almonds overnight helps to reduce their phytic acid levels, thereby enhancing the absorption of these essential minerals.

3. Peanuts

Peanuts are rich in protein, fiber, potassium, Vitamin E, and healthy fats. Just a handful (30 gm) of dry roasted peanuts can provide you with 15% of

your daily recommended vitamin E intake. Peanuts are great for maintaining a healthy heart as they help lower cholesterol levels and prevent blood clots in the arteries, which can reduce your risk of heart attack or stroke. For those with diabetes, peanuts make a perfect snack as they have a low glycemic index, which means they are slowly digested and cause a slower rise in blood sugar levels. Additionally, the manganese present in peanuts can improve glucose metabolism and increase insulin secretion.

It is recommended that you limit your daily peanut consumption to between 30 and 60 grams. While peanuts contain healthy unsaturated fats, they are also high in calories and saturated fats. To maintain normal cholesterol levels, it is important to keep your consumption of saturated fats to no more than 10% of your daily total fat intake.

VEGETABLE OILS

Adding vegetable oil to your diet is one of the easiest ways to get vitamin E, which is also a great source of healthy fats like polyunsaturated and monounsaturated fats. To get the maximum health benefits, it is important to avoid heating these oils at very high temperatures (>200 °C) and to avoid reusing the same oil multiple times. When deep frying, temperatures around 180 °C are required. However, high temperatures can cause the oil to break down and change its chemical structure, forming harmful compounds that can be absorbed by your fried food and increase the risk of cancer. To prevent this, keep the flame at medium or medium-high, and reduce it to low if you see smoke. Discard any remaining oil and avoid reusing it.

4. Wheat Germ Oil

Vitamin E is abundant in wheat germ oil. Just one tablespoon of this oil contains 135% of the daily recommended value of this important nutrient. Wheat germ oil is extracted from the germ, which is the most nutritious part of the wheat kernel. While it may have a distinct taste that some people don't find pleasant, it is incredibly nutritious. It's also rich in octacosanol, which has been shown to have anti-parkinsonism effects and can improve the way your body uses oxygen to boost muscular energy. Wheat germ oil's high vitamin E content makes it a superfood for your skin and hair. Its ability to scavenge free radicals helps prevent wrinkles and other signs of aging while also promoting good scalp health and stimulating healthy hair growth.

To extend the shelf life of your wheat germ oil, it's important to store it properly. Due to its high unsaturated fat content, it can quickly become rancid when exposed to oxygen in the air. Therefore, keeping it in an airtight container in a cool and dark location is recommended. This will help maintain the quality and freshness of the oil for a longer period of time.

5. Sunflower Oil

Sunflower oil is one of the best sources of vitamin E and can help promote heart health. Just one tablespoon of sunflower oil contains 47% of the daily recommended value of Vitamin E. The monounsaturated fat (oleic acid) found in sunflower oil is beneficial for heart health. Vitamin E helps prevent cholesterol from oxidizing and keeps blood vessels intact while also preventing clots from forming. Additionally, the high linoleic acid content found in sunflower oil can also provide cardiovascular benefits and reduce the risk of heart disease. However, it's important to consume sunflower oil in moderation, as it is high in omega-6 fats, and too much consumption can lead to inflammation.

6. Rice Bran Oil

In recent years, the popularity of rice bran oil has increased because of its potential health advantages. It contains both types of vitamin E, tocopherol and tocotrienols. Just one tablespoon of rice bran oil can provide 32% of your daily recommended vitamin E intake. It is also rich in vitamin K, monounsaturated (MUFA), and polyunsaturated fats (PUFA), as well as other vital nutrients. Studies have shown that rice bran oil has cholesterol-lowering effects. It can help lower blood pressure and is also beneficial for those with Type II diabetes. For maximum health benefits, use a blend of 80% rice bran oil with 20% sesame oil.

In addition, using rice bran oil for oil pulling can aid in fighting bad breath (halitosis). Simply take a tablespoon (15 ml) of oil into your mouth and swish it around for 10 minutes before spitting it out. Then, rinse your mouth with a glass of water for 1 minute.

Other notable oils rich in vitamin E are cottonseed oil (35% of DV), safflower oil (31% of DV), corn oil (13% of DV), and soybean oil (7% of DV).

VEGETABLES

While vegetables and fruits may not be the most abundant sources of vitamin E, they do contain some amount of this essential nutrient. Moreover, they are packed with other important nutrients and vitamins. To ensure that you meet your daily recommended intake of all vitamins, it is best to follow a balanced diet that includes grains, legumes, healthy oils, nuts, and seeds, as well as fruits and vegetables. Relying solely on one food source for your vitamin intake can deprive you of the health benefits that come from other foods.

7. Spinach

Spinach is considered a superfood due to its abundance of essential nutrients and health benefits. It contains high potassium, magnesium, calcium, iron, and vitamins A, K, and E levels. In fact, just half a cup of boiled spinach provides 13%

of the daily recommended intake of vitamin E. Additionally, spinach is known for its anti-inflammatory and anti-oxidative properties, which can help reduce the risk of high blood pressure. Studies have also found that increased consumption of spinach may lower the risk of asthma in children, likely due to its high content of beta-carotene and vitamin E, both of which have been shown to play a role in reducing airway inflammation.

Those with kidney problems should limit their consumption of spinach as consuming it in excess may cause kidney stones. Additionally, if you are taking blood-thinning medication, it is important to consult with your doctor and pharmacist to determine the appropriate daily intake of spinach, as it may reduce the effectiveness of your medication.

8. Broccoli

Do you know broccoli leaves have more antioxidants than the florets and stems? Broccoli is jam-packed with antioxidants, including vitamins E, C, and A, which help to strengthen the immune system and prevent damage to cells caused by free radicals. Plus, it's rich in sulforaphane, a powerful compound that can help prevent cancer cell formation and lower blood sugar levels. Half a cup of boiled broccoli can provide 7% of your daily recommended vitamin E intake.

Other notable vitamin E-rich vegetables are red bell peppers, pumpkin, and tomatoes.

FRUITS

9. Avocado

Avocado has a moderate amount of vitamin E. Half of an avocado (around 100 grams) contains 14% of the daily value of vitamin E, making it a healthy addition to your diet. While avocado has a high-fat content of about 75%, it's important to note that it's mostly heart-healthy monounsaturated fat, such as oleic acid. The remaining fat is

saturated fat. Interestingly, avocados contain more potassium than bananas. Consuming avocados in moderation can help protect you against cardiovascular diseases. Avocados' high monounsaturated fat content can help lower your

LDL cholesterol levels (bad cholesterol), while vitamin E can prevent cholesterol deposition by preventing blood clots and oxidative damage. Additionally, potassium can help keep your blood pressure within normal limits and reduce your heart disease and stroke risk.

10 Kiwi

Kiwis are among the richest sources of vitamin C and have a generous amount of antioxidant vitamin E, which are antioxidants that can help boost your immunity and prevent chronic inflammatory diseases. Just one medium green kiwi can provide 7% of your daily requirement of vitamin E . These antioxidants work by killing free radicals before they can cause damage to your cells. Kiwis are also rich in potassium, which helps to promote blood vessel relaxation and lower blood pressure. Plus, the fiber in kiwis can help prevent constipation and keep your cholesterol levels in check.

Other notable fruits rich in vitamin E are mango and blackberry.

Effect of Cooking Temperature on Vitamin E:

Vitamin E in Green Vegetables

Boiling or blanching your green leafy vegetables can increase their vitamin E content. This is because cooking breaks down the cell walls of plants, which releases vitamin E from the lipids. However, cutting or mixing vegetables can activate oxidizing enzymes that cause vitamin E loss. To prevent this, it's best to use heat treatment, which deactivates these enzymes and preserves the vitamin E content. Therefore, it's best to boil or blanch green vegetables instead of eating them raw to get maximum vitamin E.

Vitamin E in Oils

When cooking with Vitamin E, it's important to be mindful of the temperature and cooking time. While Vitamin E is generally stable in heat, high-heat cooking methods like frying can cause a loss of this important nutrient. Studies have shown that heating oil at 210 °C or more can result in a loss of 6.38% of Vitamin E. Additionally, cooking at very high temperatures (over 250 °C) for longer periods of time can reduce the Vitamin E content by up to 20%. To ensure you're getting the most out of Vitamin E-rich foods like almonds and cooking oils, it's best to avoid deep frying or baking at high temperatures (over 200 °C) and cook them for shorter periods of time.

CHAPTER 5

POTENTIALLY DANGEROUS VITAMIN E COMBINATIONS

When taking vitamins, it's important to consider how different combinations may affect the body. For vitamins to produce their intended effects, they must be adequately absorbed in the body. Combining certain vitamins with other vitamins, medications, or minerals can either enhance or hinder their absorption rates. Some vitamin combinations may have a synergistic effect, increasing the absorption of other vitamins and providing better health benefits than if taken alone. However, other combinations may compete for absorption in the body, nullifying their effects and potentially causing toxicity.

VITAMIN E + VITAMIN K

Vitamin K is important for maintaining bone health, promoting wound healing, and aiding blood clotting. However, when taken with vitamin E, the effects of Vitamin K can be reduced. Vitamin E has blood thinning effects and also increases the metabolism of Vitamin K in the liver, which leads to increased excretion of all forms of vitamin K. While consuming vitamin K-rich foods like spinach and kale with vitamin E-rich foods like nuts or vegetable oil in moderation may not have adverse effects, high intake of vitamin E-rich foods or taking Vitamin E supplements alongside Vitamin K supplements can diminish the benefits of Vitamin K in the body.

Vitamin E + WATER-SOLUBLE VITAMINS

In order to receive the health benefits of vitamins, they must be properly absorbed. It's not recommended to take fat-soluble vitamins (A, D, E, and K) with water-soluble vitamins (B complex and C) because they are absorbed differently in the body. Combining them may reduce the health benefits you receive from each. Water-soluble vitamins are well absorbed on an empty stomach, while fat-soluble vitamins require the presence of fat in the body to be adequately absorbed. To maximize the benefits of each type of vitamin, consume B and C-rich foods in the morning and foods rich in A, D, E, and K in the evening. If you take vitamin supplements, take B and C on an empty stomach and take fat-soluble vitamins in the evening after a meal.

CHAPTER 6

HEALTHY VITAMIN E COMBINATIONS

VITAMIN E + OMEGA-3 FATS

Reducing inflammation in the body is important to prevent chronic disorders such as hypertension, type 2 diabetes, kidney disease, and cardiovascular disease. Omega-3 fatty acids help in reducing inflammation.

Vitamin E has potent antioxidant effects, which protect cells from damage caused by free radicals and limit their harmful effects. Vitamin E also provides protection against cancer and cardiovascular diseases while enhancing the immune system.

Omega-3 fatty acids, such as alpha-linolenic acid (ALA), eicosapentaenoic acid (EPA), and docosahexaenoic acid (DHA), are polyunsaturated (PUFA) and are prone to oxidation in the gut. Omega-3-rich foods like flaxseeds, chia seeds, and canola oil can even be oxidized during processing and storage, which reduces their stability and health benefits. Oxidized Omega-3 fatty acids cause oxidative stress and have deteriorating effects on the body. However, combining antioxidants like vitamin E with Omega 3 can prevent its oxidation and increase its stability in the gut.

For maximum health benefits, it's recommended to consume foods that are high in vitamin E, like sunflower seeds, almonds, peanuts, pumpkin, spinach, and mangoes, with omega-3 fats-rich foods, such as flaxseeds, chia seeds, walnuts, and kidney beans.

VITAMIN E + VITAMIN C

Both vitamin E and vitamin C have antioxidant properties, but they differ in solubility. Vitamin C is a water-soluble antioxidant, while vitamin E is a lipid-soluble antioxidant. When combined, these two vitamins can boost immunity, reduce the risk of developing chronic diseases, and protect against exercise-induced damage.

Vitamin E is a potent antioxidant. However, when combined with Vitamin C, it becomes even more effective. Vitamin E's main role is to act as a lipid antioxidant, which helps protect the polyunsaturated fatty acids (PUFAs) found in the membrane from lipid peroxidation. When free radicals attack these PUFAs, it can lead to cell damage. Vitamin E can prevent this by binding with free radicals and rendering them ineffective. However, this process also causes Vitamin E to change into a form that can no longer bind with other free radicals. Luckily, Vitamin C can solve this problem.

Vitamin C helps recycle vitamin E. Vitamin C enhances vitamin E's antioxidant activity, making it more efficient. After vitamin E binds with free radicals and becomes oxidized, vitamin C steps in and reduces it back to its original form of tocopherol. With vitamin E restored, it can bind with free radicals again, rendering them ineffective. Together, vitamin C and vitamin E form an antioxidant network that protects against chronic diseases by safeguarding lipids, proteins, and DNA against free radical damage.

Studies indicate that consuming foods that are rich in vitamins E and C can lower the risk of reduced lung function or asthma, particularly in people who are at an increased risk of developing these conditions, such as smokers. However, taking supplements of both vitamins did not yield the same results. The combination of vitamins E and C supplements has produced mixed results, and further studies are needed to draw a definitive conclusion.

For maximum health benefits, incorporate vitamin C-rich foods like citrus fruits, strawberries, and tomatoes, as well as vitamin E foods such as almonds, sunflower oil, pumpkin, and red bell peppers into your diet. However, it is not advisable to consume water-soluble vitamins like vitamin C with fat-soluble vitamins. Therefore, it is recommended to maintain a gap of 2-3 hours between the consumption of both types of vitamins.

VITAMIN E + SELENIUM

Research has shown that a combination of vitamin E and selenium has a synergetic effect, making it more effective in protecting against prostate cancer and atherosclerosis compared to using either of these alone.

Selenium acts as both an immunomodulator and an antioxidant. It is even more powerful than vitamins E, C, and A in terms of its antioxidant properties.

Both vitamin E and selenium are antioxidants that play a crucial role in protecting cells. They safeguard lipids, specifically polyunsaturated fatty acids, present in cell membranes from oxidative degradation caused by free radicals. If antioxidants do not shield these lipids, lipid peroxidation can occur, leading to the destruction of membrane lipids. This can be harmful to the viability of cells and tissues and can make you more susceptible to chronic diseases such as atherosclerosis, inflammatory bowel disease, asthma, Parkinson's disease, and kidney damage.

Both selenium and vitamin E are antioxidants, but they function differently. Selenium increases the reactivity of an enzyme called glutathione peroxidase (GPx), which helps protect the body from oxidative damage. On the other hand, vitamin E is a chain-breaking antioxidant that interrupts the chain reaction by pairing with free radicals and preventing them from pairing with lipids.

It is important to obtain selenium from your diet, as this mineral is highly potent in antioxidants but can also be highly toxic. To ensure a balanced intake, try incorporating vitamin E-rich foods such as almonds, peanuts, asparagus, spinach, and Swiss chard with selenium-rich foods like Brazil nuts, walnuts, sunflower seeds, chia seeds, flax seeds, broccoli, garlic, onion, cottage cheese, mushrooms, and brown rice.

CHAPTER 7

DIET PLAN

Here's a 10-day diet plan to include natural sources rich in vitamin E in your diet. Repeat this diet plan every 10 days and you will never be deficient in Vitamin E.

Day 1	1 tablespoon wheat germ oil (>100%)
Day 2	A handful of dry roasted hazelnuts (30%) + 1 medium Kiwi (10%) + 2 tablespoons of safflower oil (60%)
Day 3	2 tablespoons of peanut butter (20%) + A handful of dry roasted hazelnuts (30%) + 1 cup cooked spinach (30%) + 2 medium kiwi (20%)
Day 4	2 tablespoons of sunflower oil (80%) + 40 g of dry roasted peanuts (20%)
Day 5	A handful of sunflower seeds and a handful of roasted almonds (95%) + 1

	tomato
Day 6	1 cup spinach cooked in 1½ tablespoons of sunflower oil (85%) + A handful of peanuts (15%)
Day 7	1 cup spinach cooked in 1½ tablespoons of sunflower oil (85%) + A handful of peanuts (15%)
Day 8	A handful of dry roasted sunflower seeds (45%) + ¾ cup broccoli cooked in 1 tablespoon of sunflower oil (55%)
Day 9	2 tablespoons corn oil (30%) + 1 cup cooked spinach (30%), 1 large avocado (25%) +3 tomatoes (15%)
Day 10	2 tablespoons peanut butter (20%)+ a handful of roasted almonds (45%)+ 2 kiwi (15%)

CHAPTER 8

RECIPES

Healthy and delicious vitamin E vegetarian recipes for maximum health benefits.

Stir Fried Broccoli

Ingredients

Broccoli: 1	Garlic: 4
Peanuts: 2 tbsp	Asafetida: ¼ tsp
Blackpepper powder: ¼ tsp	Salt: To taste
Red chili flakes: a pinch	Sunflower oil: 1 tbsp

Method

1. Cut the broccoli into 2-inch pieces. Sprinkle salt and steam them in a steamer or pressure cooker, or para boil them in a saucepan.

2. Dry roast the peanuts. Place peanuts in a kitchen towel, cover them with a kitchen towel, and rub them to remove the skin. Crush the peanuts in a mortar.

4. Heat oil and add asafetida and chopped garlic to it. Cook them till they turn golden. Add peanuts and cook for 2 minutes.

6. Add broccoli, black pepper powder and chili flakes. Stir fry for 5 minutes and it's ready to eat.

Peas Avocado Soup

Ingredients

Avocado: 1 (170 g)	Peas: 100 g
White onion: 1 medium	Garlic: 5-6 cloves
Cumin seeds: ¼ tsp	Nutmeg: ½ small / ¼ tsp
Garam masala: a pinch / ¼ tsp	Black pepper powder: ¼ tsp
Curd: 2 tsp	Salt: To taste
Rice bran oil: 1 tbsp	Water 400 ml
Optional topping: ¼ tsp tandoori masala	

Method

1. Heat oil in a saucepan. Add cumin seeds. When cumin starts crackling, add garlic and cook for 2 minutes.

2. Add chopped onion and cook till it turns slightly brown.

3. Add fresh peas and salt. Cover and cook till the peas become soft.

4. Lower the flame and add curd. Mix well and cook covered for 2 minutes.

5. Add a pinch of garam masala, black pepper powder and grate about half a small nutmeg. Cover and cook for 2 minutes.

6. Add 200 ml water. Cover and cook for 5 minutes till the water reduces slightly.

7. Add the remaining 200 ml water and cover again and cook for 5 minutes till the water reduces slightly.

8. Take a hand blender and blend the soup to make it thick and smooth.

9. Turn off the flame and add chopped avocado to it. Blend all the ingredients with a hand blender to make a creamy smooth pea avocado soup.

10. Take out the soup in a bowl and optionally sprinkle tandoori masala over it for a tangy taste.

Kiwi Smoothie

Ingredients

Kiwi: 2	Papaya: 150 g / 2 slices
Curd: 100 g	Chia seeds: 1 tbsp
Ginger: ¼ inch	Black salt: To taste
Water/Coconut water: 100 ml	

Method

1. Soak chia seeds in plain water or coconut water for 2 hours.

2. Add kiwi, curd, papaya and black salt in a blender jar. Grate ginger and blend everything to make a smoothie.

3. Lastly add chia seeds and blend one last time to make a smooth kiwi smoothie.

Almond Avocado Pudding

Ingredients

Almond: 50 g	Honey: 2 tbsp
Cocoa powder: 2 tbsp	Olive oil: 1 tbsp
Avocado: 1	

Method

1. Dry roast the almonds till they change color.
2. Blend the almonds in a food processor until they release oil. Once it turns into almond butter, add honey and oil and blend again.
3. Add finely chopped avocado and blend until the avocado is mixed with the almonds.
4. Lastly, add cocoa powder and blend to make a smooth, lump-free Almond Avocado Pudding.

The End

Sign up to La Fonceur Newsletter to receive Bonus Recipes:

http://eatsowhat.com/signup

REFERENCES

1. Multivitamin/Mineral Supplements Fact Sheet for Health Professionals, National Institutes of Health.
2. Blaner WS, Shmarakov IO, Traber MG. Vitamin A and Vitamin E: Will the Real Antioxidant Please Stand Up? Annu Rev Nutr. 2021 Oct 11;41:105-131. doi: 10.1146/annurev-nutr-082018-124228. Epub 2021 Jun 11.
3. Vitamin E Health Professionals Fact Sheet, National Institutes of Health.
4. Kemnic TR, Coleman M. Vitamin E Deficiency. [Updated 2023 Jul 4]. In: StatPearls [Internet]. Treasure Island (FL): StatPearls Publishing; 2023 Jan.
5. Vitamin E (Tocopherol) Test - MedlinePlus - National Library of Medicine.
6. Uchiyama K, Kishi H, Komatsu W, Nagao M, Ohhira S, Kobashi G. Lipid and Bile Acid Dysmetabolism in Crohn's Disease. J Immunol Res. 2018 Oct 1;2018:7270486.
7. Kretzer FL, Mehta RS, Johnson AT, Hunter DG, Brown ES, Hittner HM. Vitamin E protects against retinopathy of prematurity through action on spindle cells. Nature. 1984 Jun 28-Jul 4;309(5971):793-5. doi: 10.1038/309793a0.
8. Effect of Vitamin E for Prevention of Retinopathy of Prematurity: A Randomized Clinical Trial.

9. Rosca MG, Vazquez EJ, Kern TS, Hoppel CL. Oxidation of fatty acids is source of increased mitochondrial reactive oxygen species production in kidney cortical tubules in early diabetes. Diabetes. 2012 Aug;61(8):2074-83. Epub 2012 May 14.
10. Aprioku JS. Pharmacology of free radicals and the impact of reactive oxygen species on the testis. J Reprod Infertil. 2013 Oct;14(4):158-72.
11. Bardaweel SK, Gul M, Alzweiri M, Ishaqat A, ALSalamat HA, Bashatwah RM. Reactive Oxygen Species: the Dual Role in Physiological and Pathological Conditions of the Human Body. Eurasian J Med. 2018 Oct;50(3):193-201.
12. S. Nazrun, M. Norazlina, M. Norliza, S. Ima Nirwana, "The Anti-Inflammatory Role of Vitamin E in Prevention of Osteoporosis", Advances in Pharmacological and Pharmaceutical Sciences, vol. 2012, Article ID 142702, 7 pages, 2012.
13. Zhang JM, An J. Cytokines, inflammation, and pain. Int Anesthesiol Clin. 2007 Spring;45(2):27-37.
14. Konopatskaya O, Matthews SA, Poole AW. Protein kinase C mediates platelet secretion and thrombus formation through protein kinase D2. Blood. 2011 Jul 14;118(2):416-24.
15. Kunisaki M, Umeda F, Inoguchi T, Nawata H. Vitamin E restores reduced prostacyclin synthesis in aortic endothelial cells cultured with a high concentration of glucose.

Metabolism. 1992 Jun;41(6):613-21. doi: 10.1016/0026-0495(92)90053-d.

16. Wu D, Liu L, Meydani M, Meydani SN. Effect of vitamin E on prostacyclin (PGI2) and prostaglandin (PG) E2 production by human aorta endothelial cells: mechanism of action. Ann N Y Acad Sci. 2004 Dec;1031:425-7. doi: 10.1196/annals.1331.063.
17. Dayong Wu, Liping Liu, Mohsen Meydani, Simin Nikbin Meydani, Vitamin E Increases Production of Vasodilator Prostanoids in Human Aortic Endothelial Cells through Opposing Effects on Cyclooxygenase-2 and Phospholipase A2, The Journal of Nutrition, Volume 135, Issue 8, August 2005, Pages 1847–1853.
18. Lewis ED, Meydani SN, Wu D. Regulatory role of vitamin E in the immune system and inflammation. IUBMB Life. 2019 Apr;71(4):487-494. doi: 10.1002/iub.1976. Epub 2018 Nov 30.
19. Cologne, Germany: Institute for Quality and Efficiency in Health Care (IQWiG); 2006-. What is an inflammation? 2010 Nov 23 [Updated 2018 Feb 22].
20. Lee GY, Han SN. The Role of Vitamin E in Immunity. Nutrients. 2018 Nov 1;10(11):1614. doi: 10.3390/nu10111614.
21. Jiang T, Zhou C, Ren S. Role of IL-2 in cancer immunotherapy. Oncoimmunology. 2016 Apr 25;5(6):e1163462.

22. Coronary Heart Disease - National Health Service UK.
23. Waters DD, Alderman EL, Hsia J, Verter JI. Effects of hormone replacement therapy and antioxidant vitamin supplements on coronary atherosclerosis in postmenopausal women: a randomized controlled trial. JAMA. 2002 Nov 20;288(19):2432-40. doi: 10.1001/jama.288.19.2432.
24. Lee IM, Cook NR, Gaziano JM, Gordon D, Ridker PM, Manson JE, Hennekens CH, Buring JE. Vitamin E in the primary prevention of cardiovascular disease and cancer: the Women's Health Study: a randomized controlled trial. JAMA. 2005 Jul 6;294(1):56-65. doi: 10.1001/jama.294.1.56.
25. Sesso HD, Gaziano JM. Vitamins E and C in the prevention of cardiovascular disease in men: the Physicians' Health Study II randomized controlled trial. JAMA. 2008 Nov 12;300(18):2123.
26. Alkhenizan A, Hafez K. The role of vitamin E in the prevention of cancer: a meta-analysis of randomized controlled trials. Ann Saudi Med. 2007 Nov-Dec;27(6):409-14.
27. Scanlan RA. Formation and occurrence of nitrosamines in food. Cancer Res. 1983 May;43(5 Suppl):2435s-2440s.
28. Gugliandolo A, Mazzon E. Role of Vitamin E in Treatment of Alzheimer's Disease: Evidence

from Animal Models. Int J Mol Sci. 2017 Nov 23;18(12):2504. doi: 10.3390/ijms18122504.
29. Zhang SM, Hernan MA, Chen H, Spiegelman D, Willett WC, Ascherio A. Intakes of vitamins E and C, carotenoids, vitamin supplements, and PD risk. Neurology. 2002;59:1161-9.
30. Keen MA, Hassan I. Vitamin E in dermatology. Indian Dermatol Online J. 2016 Jul-Aug;7(4):311-5.
31. Cook-Mills JM, Avila PC. Vitamin E and D regulation of allergic asthma immunopathogenesis. Int Immunopharmacol. 2014 Nov;23(1):364-72. doi: 10.1016/j.intimp.2014.08.007. Epub 2014 Aug 29.
32. Dadkhah H, Ebrahimi E, Fathizadeh N. Evaluating the effects of vitamin D and vitamin E supplement on premenstrual syndrome: A randomized, double-blind, controlled trial. Iran J Nurs Midwifery Res. 2016 Mar-Apr;21(2):159-64. doi: 10.4103/1735-9066.178237.
33. Ronald J. Sokol, Maret G. Traber. Vitamin E and Vitamin K Metabolism. Physiology of the Gastrointestinal Tract (Fourth Edition), 2006.
34. Traber MG. Vitamin E and K interactions--a 50-year-old problem. Nutr Rev. 2008 Nov;66(11):624-9. doi: 10.1111/j.1753-4887.2008.00123.x.
35. Podszun M, Frank J. Vitamin E-drug interactions: molecular basis and clinical relevance. Nutr Res Rev. 2014 Dec;27(2):215-

31. Doi: 10.1017/S0954422414000146. Epub 2014 Sep 16.

36. Kim JM, White RH. Effect of Vitamin E on the anticoagulant response to warfarin. Am J Cardiol. 1996 Mar 1;77(7):545-6. doi: 10.1016/s0002-9149(97)89357-5.
37. Drug Interactions between vitamin E and Warfarin - Professional from Drugs.com; c1996-2018.
38. Saboori S, Djalali M, Nematipour E, Ramezani A. Various Effects of Omega 3 and Omega 3 Plus Vitamin E Supplementations on Serum Glucose Level and Insulin Resistance in Patients with Coronary Artery Disease. Iran J Public Health. 2016 Nov;45(11):1465-1472.
39. M. Sepidarkish, M. Morvaridzadeh, J. Heshmati. Effect of omega-3 fatty acid and vitamin E Co-Supplementation on lipid profile: a systematic review and meta-analysis. Diabetes Metab. Syndr. Clin. Res. Rev., 13 (2019), pp. 1649-1656
40. Lu, T.; Shen, Y.; Wang, J.H.; Xie, H.K.; Wang, Y.F.; Zhao, Q.; Zhou, D.-Y.; Shahidi, F. Improving oxidative stability of flaxseed oil with a mixture of antioxidants. J. Food Proc. Preserv. 2020, 44, e14355.
41. Floros S, Toskas A, Vareltzis P. Bioaccessibility, Oxidative Stability of Omega-3 Fatty Acids in Supplements, Sardines and Enriched Eggs Studied Using a Static In Vitro

Gastrointestinal Model. Molecules. 2022 Jan 9;27(2):415. doi: 10.3390/molecules27020415.
42. Institute of Medicine (US) Committee on Military Nutrition Research. Military Strategies for Sustainment of Nutrition and Immune Function in the Field. Washington (DC): National Academies Press (US); 1999. 13, Vitamin E, Vitamin C, and Immune Response: Recent Advances.
43. Traber MG, Stevens JF. Vitamins C and E: beneficial effects from a mechanistic perspective. Free Radic Biol Med. 2011 Sep 1;51(5):1000-13. doi: 10.1016/j.freeradbiomed.2011.05.017.
44. X. Chen, Rhian M. Touyz, J.B Park. Schiffrin. Antioxidant Effects of Vitamins C and E Are Associated with Activation of Vascular NADPH Oxidase and Superoxide Dismutase in Stroke-Prone SHR. Originally published1 Sep 2001.
45. Mylonas C, Kouretas D. Lipid peroxidation and tissue damage. In Vivo. 1999 May-Jun;13(3):295-309. PMID: 10459507.
46. Noaman E, Zahran AM, Kamal AM, Omran MF. Vitamin E and selenium administration as a modulator of antioxidant defense system: biochemical assessment and modification. Biol Trace Elem Res. 2002 Apr;86(1):55-64. doi: 10.1385/BTER:86:1:55.
47. Reagan-Shaw S, Nihal M, Ahsan H, Mukhtar H, Ahmad N. Combination of vitamin E and selenium causes induction of apoptosis of

human prostate cancer cells by enhancing Bax/Bcl-2 ratio. Prostate. 2008 Nov 1;68(15):1624-34. doi: 10.1002/pros.20824.

48. Dawn C. Schwenke and Stephen R. Behr. Vitamin E Combined with Selenium Inhibits Atherosclerosis in Hypercholesterolemic Rabbits Independently of Effects on Plasma Cholesterol Concentrations Originally. 24 Aug 1998.
49. Tinggi U. Selenium: its role as antioxidant in human health. Environ Health Prev Med. 2008 Mar;13(2):102-8. doi: 10.1007/s12199-007-0019-4. Epub 2008 Feb 28.
50. Omega 3 fatty acids fact sheet – Health Professional Fact Sheet. National Institutes of Health Office of Dietary Supplements.

ABOUT THE AUTHOR

With a Master's Degree in Pharmacy, the author La Fonceur is a Research Scientist and Registered Pharmacist. She specialized in Pharmaceutical Technology and worked as a research scientist in the pharmaceutical research and development department. She is a health blogger and a dance artist. Her previous books include Eat to Prevent and Control Disease, Secret of Healthy Hair, and Eat So What! series. Being a research scientist, she has worked closely with drugs and based on her experience, she believes that one can prevent most of the diseases with nutritious vegetarian foods and a healthy lifestyle.

READ MORE FROM LA FONCEUR

English Editions

Hindi Editions

CONNECT WITH LA FONCEUR

Instagram: @la_fonceur | @eatsowhat

Facebook: LaFonceur | eatsowhatblog

Twitter: @la_fonceur

Follow on Bookbub: @eatsowhat

Sign up to get exclusive offers on La Fonceur books:

Blog: http://www.eatsowhat.com/

Website: http://www.lafonceurbooks.com/

9 789354 087929

Printed by Libri Plureos GmbH in Hamburg,
Germany